Sabrina MEJDOUB

Determination of thyroglobulin and anti-thyroglobulin antibodies

Sabrina MEJDOUB

Determination of thyroglobulin and anti-thyroglobulin antibodies

Indications, technical aspects and interpretation

ScienciaScripts

Imprint

Cover image: www.ingimage.com

This book is a translation from the original published under ISBN 978-620-6-72647-0.

Publisher:
Sciencia Scripts
is a trademark of
Dodo Books Indian Ocean Ltd. and OmniScriptum S.R.L publishing group

120 High Road, East Finchley, London, N2 9ED, United Kingdom
Str. Armeneasca 28/1, office 1, Chisinau MD-2012, Republic of Moldova, Europe
Managing Directors: Ieva Konstantinova, Victoria Ursu
info@omniscriptum.com

Printed at: see last page
ISBN: 978-620-8-87323-3

TABLE OF CONTENTS

List of abbreviations

Ac	Antibodies
Anti-TG antibodies	Anti-thyroglobulin antibodies
Ag	Antigen
DO	Optical density
ECL	Electrochemiluminescence
ELISA	Enzyme linked immunosorbent assay
T3	Tri-iodothyronine
T4	Tetra-iodothyronine
TG	Thyroglobulin
TMB	3,3′, 5,5′-tetramethylbenzidine
TPA	Tripropylamine
TRH	Thyrotropin releasing hormone
TSH	Thyroid stimulating hormone

I. Theoretical

1. The thyroid gland

1.1. Anatomy :

The thyroid gland is an impolar median gland located on the anterior surface of the neck, in front of the laryngotracheal axis. It consists of two lobes joined by an isthmus.

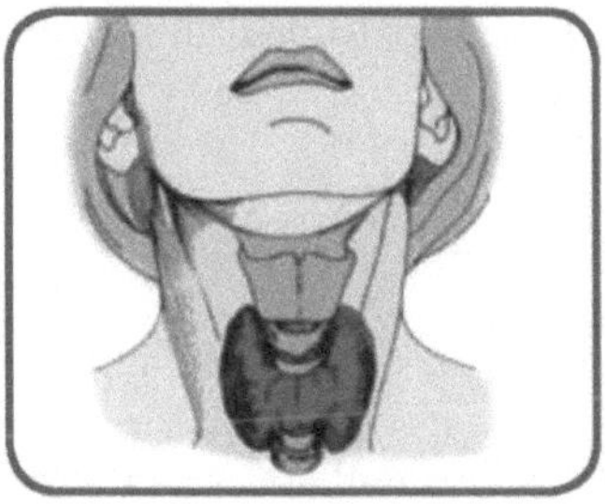

Figure 1: Anterior view of the neck showing the location of the thyroid.

1.2. Histology :

The thyroid follicle (or thyroid vesicle) represents the morpho-functional unit of the thyroid. It has a lining of unistratified epithelium, and a central cavity containing the colloid.

- The follicular wall comprises two types of epithelial cells:
  - follicular cells or vesicular cells or thyreocytes, which secrete thyroid hormones (cells at the origin of differentiated thyroid cancers)
  - parafollicular cells or C cells, secreting calcitonin (cells at the origin of medullary thyroid cancers).
- The colloid consists mainly of a 660 kDa glycoprotein: thyroglobulin (TG).

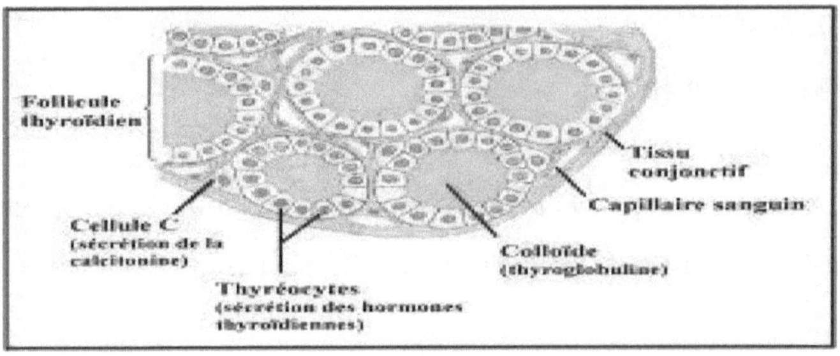

Figure 2: Diagram of a histological section of the thyroid.

1.3. Physiology :

➢ **<u>Thyroid hormone synthesis</u> :**

The synthesis of thyroid hormones (T3 and T4) involves several stages:

- Capture of blood iodine at the basal pole of the thyreocyte
- Oxidation of iodine captured in the thyreocyte and its concentration in the colloid
- TG synthesis :

TG is a glycoprotein synthesized by thyreocytes. Its protein fraction is synthesized by ribosomes in the granular endoplasmic reticulum from blood amino acids (tyrosine). After glycosylation in the Golgi apparatus, TG is released into the colloid by exocytosis.

- Incorporation of iodine into the colloid at TG tyrosine residues in the form of mono-iodotyrosines (MIT) and di-iodotyrosines (DIT).
- Condensation of MIT and DIT into tri-iodothyronine (T3) and tetra-iodothyronine (T4)
- Colloid resorption in the thyrocyte by endocytosis
- Proteolysis of TG and release of T3 and T4 hormones
- Excretion of T3 and T4 into blood capillaries located in interstitial tissue

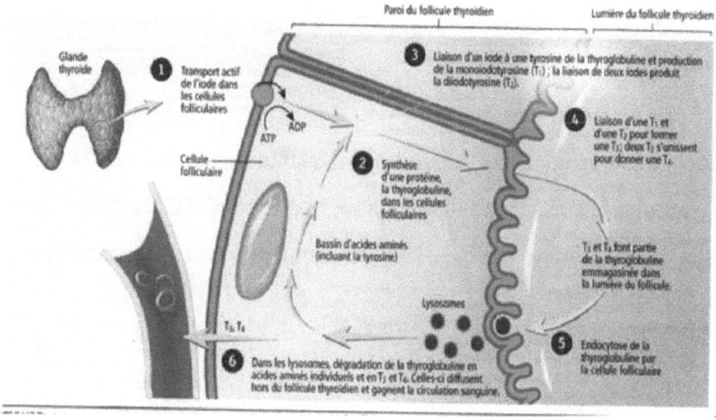

Figure 3: Stages of thyroid hormone synthesis

➢ Regulation of thyroid hormone synthesis :

Thyroid function is regulated by the hypothalamus-pituitary axis. The hypothalamus secretes TRH, which activates TSH secretion by the pituitary, which in turn stimulates synthesis and release of T3 and T4 by the thyroid.

By negative feedback, T3 and T4 inhibit the production of TSH and TRH.

TSH regulates the various stages of thyroid hormone synthesis. It also acts as a growth factor for the thyroid gland.

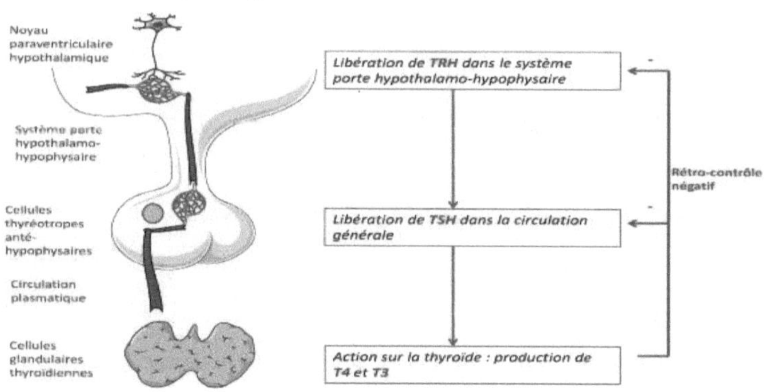

Figure 4: Hypothalamic-pituitary regulation of thyroid secretion

2. Differentiated thyroid cancer

2.1. Definition :

Differentiated thyroid cancers are epithelial malignant tumors which develop from thyreocytes, and which have retained several characteristics of these cells (TSH sensitivity, iodine uptake, TG secretion). These cancers generally have a good prognosis.

2.2. Epidemiology :

While thyroid nodules are common, thyroid cancers are relatively rare. However, their incidence has been rising for several decades, thanks to improved practices for early detection.

Differentiated thyroid cancers account for over 85% of all thyroid cancers. Anatomopathologically, they are classified into papillary cancers, the most frequent (70%), and follicular or vesicular cancers.

They predominate in young adults (median age at diagnosis 45-50), and are more common in women than in men. Cervical irradiation in childhood is the only clearly established risk factor.

2.3. Diagnosis :

➢ **Interrogatory and clinical examination**:

Most patients present with an asymptomatic thyroid nodule. Most often, the nodule is palpable, painless, firm and mobile when swallowed. Occasionally, the nodule is not palpable and is discovered incidentally by an ultrasound scan performed for another reason.

Signs of strong suspicion of malignancy are sometimes found: very firm or hard, irregular, fixed nodule (adherent to surrounding tissues), increase in size, presence of compressive signs (dysphonia, cough, dyspnea, dysphagia), presence of cervical adenopathies.

More rarely, differentiated thyroid cancer may be discovered through cervical lymph node metastases or distant metastases, notably in the lungs or bones.

➢ **Cervical ultrasound :**

Ultrasound clarifies the characteristics of the nodule (making it possible to assess the risk of cancer for each nodule), looks for other abnormalities in the thyroid and cervical lymph node chains, and guides cytopuncture. Nodules measuring 1 cm or less, with no suspicion of cancer, are monitored.

➢ **Fine needle aspiration:**

Cytopuncture is performed on any solid nodule measuring more than 1cm or presenting elements of suspicion of malignancy.

There are four possible results: benign (70%), malignant (5%), suspicious (10%) and non-informative (cytopuncture did not yield sufficient cells) (15%).

Malignant and suspicious results lead to thyroidectomy. Non-informative nodules should be resected without delay. So-called benign nodules will be re-sponctioned once, within a year.

➢ **Biology :**

- **TSH** :

Virtually all thyroid cancer patients are euthyroid; TSH levels are generally normal. Thyroid cancers synthesize little or no thyroid hormone (i.e. no hyperthyroidism), and very rarely destroy all normal thyroid parenchyma (i.e. no hypothyroidism). However, it is possible for a patient to have both thyroid cancer and dysthyroidism.

- **TG**:

Differentiated thyroid carcinomas secrete TG, but no more than normal thyreocytes. **TG assay is of no value in diagnosing the malignancy of a nodule.**

2.4. Treatment :

➢ **Surgical thyroidectomy:**

Surgery is the main treatment for differentiated thyroid cancer. Its aim is to remove all cervical cancerous tissue. The procedure consists of a more or less total thyroidectomy, with or without lymph node dissection.

➢ **Isotopic treatment with radioactive iodine 131 :**

When a total thyroidectomy has been performed, the procedure is completed by isotopic destruction (totalization) with iodine 131. This irradiation is generally carried out one month after surgery, without the patient having received any replacement therapy. At this point, the patient is hypothyroid with elevated TSH, which stimulates iodine uptake by cervical residues and any metastases. The administration of iodine-131 therefore reveals and destroys :

- normal thyroid remnants (ablation), thus facilitating subsequent monitoring by TG assay,
- local tumour residues and distant metastases undetected before surgery (treatment), thus reducing the relapse rate.

Treatment with iodine-131 also enables a whole-body scan to be carried out three to seven days later, verifying the absence of fixation foci outside the thyroid area.

➢ **Medical treatment with levothyroxine :**

L-thyroxine treatment is prescribed for life to all patients undergoing surgery for differentiated thyroid cancer (started two to three days after administration of iodine 131 at). The aim is to administer LT4, not just at a replacement dose that replaces thyroid secretion after thyroidectomy, but at a dose sufficient to inhibit TSH secretion to prevent the growth of any micro-metastases.

2.5. Follow-up :

As **TG** is only produced by thyroid cells of follicular origin, its measurement is an **excellent marker for follow-up after total thyroidectomy (followed by isotopic destruction of the remnants).**

For the TG assay to be interpretable, it must be ensured that there are no anti-TG antibodies (Ac) (source of interference leading to underestimated results/false negatives). Anti-TG antibodies must therefore be systematically requested at the same time as the TG assay.

- Initially, TG is measured **on the day of iodine-131 administration**. An undetectable level has an excellent predictive value for the absence of residual disease.

  A whole-body scan is performed three to seven days later to verify the absence of fixation foci outside the thyroid area.

- 3 months after the start of L-thyroxine treatment, TSH levels are measured to check hormonal balance.

- **6 to 12 months after the initial treatment**, a healing check-up is carried out, including cervical ultrasound and TG measurement after thyroid hormone withdrawal or stimulation with recombinant human TSH (intra-muscular injection, two days in a row, then TG measurement three days after the second injection).

  In over 80% of patients, both tests are normal, and these patients are considered cured. The dose of L-thyroxine is reduced to obtain a TSH level within the normal range.

- Subsequent monitoring is **annual**, with TSH and TG measurements taken during L-thyroxine treatment.

  In some patients, late relapses may occur, so monitoring is continued **for life**.

Evolution of TG in the event of relapse :

During LT4 treatment, TG levels are elevated in most relapsed patients. However, it is undetectable in over 20% of patients with isolated lymph node metastases, and in 5% of patients with sub-radiological lung metastases.

After thyroid hormone withdrawal or stimulation with recombinant human TSH, TG levels are detectable in the majority of relapsed patients. False-negatives are associated with small lymph node metastases detected by cervical ultrasound. Further investigations are then undertaken. Depending on the nature of the relapse (cervical or distant), its location, extension and whether or not it is fixative, surgical treatment and/or internal or external radiotherapy are indicated.

II. Aim of the study

In this study, we addressed two aspects:

- Determination of circulating tumour markers, specifically TG
- The detection of auto-Ac, specifically anti-TG Ac.

Objective: To describe the methodology and interpret the results of the TG and anti-TG Ac assays in the follow-up of differentiated thyroid cancer.

III. Materials and Methods

1. Pre-analytical phase :

"Series of steps beginning chronologically with the clinician's prescription for analyses, including: the request for analysis, preparation of the patient, specimen collection, transport to and within the laboratory and ending at the start of the analytical procedure."

➢ Patient preparation :

- Generally, no special preparation is required. Fasting is not required.

- In patients receiving high-dose biotin therapy (>5 mg/day), sampling for TG determination should be performed at least 8 hours after the last biotin administration.

➢ Sampling :

- A venous blood sample is taken. This involves puncturing a superficial vein with a suitable needle to collect a venous blood sample in a collection tube.

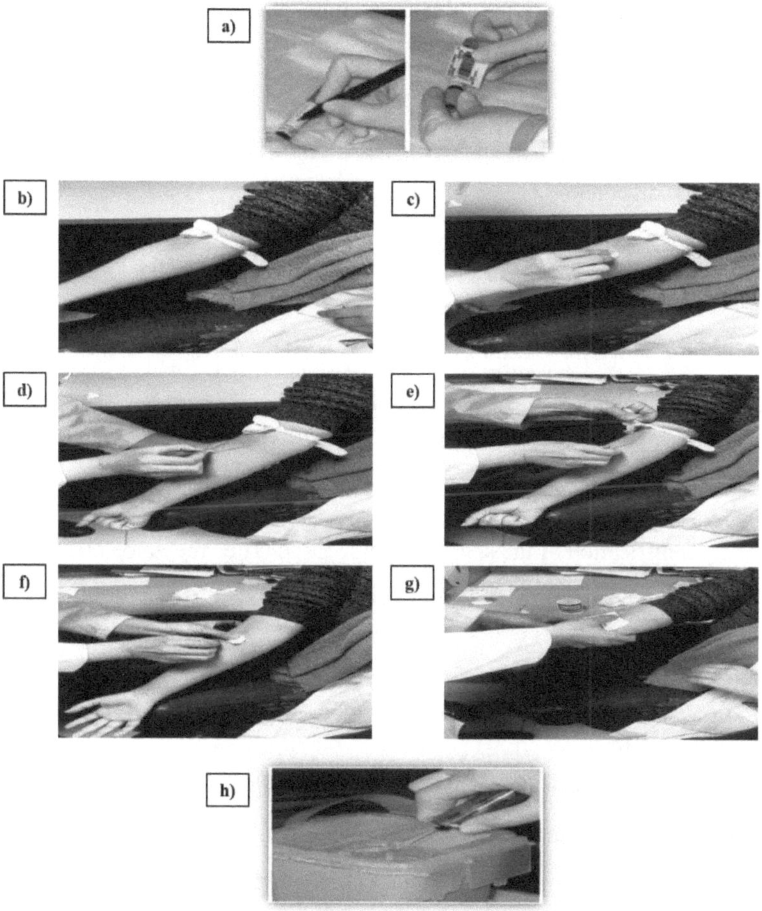

Figure 5: Sampling procedure

a) Step 0 "Patient label
b) Step 1 "Fix the tourniquet
c) Step 2 "Select vein and disinfect puncture site".
d) Step 3 "point the bevel of the needle upwards. Penetrate the vein at an angle of 30° or less".
e) Step 4 "as soon as blood flows, release the tourniquet and ask the patient to open his hand".
f) Step 5 "after removing the tube, cover the puncture site with gauze and withdraw the needle slowly, maintaining the angle of insertion".
g) Step 6 "apply firm pressure to the vein for 15 to 30 seconds and dress".
h) Step 7 "dispose of the needle in a suitable container".

- The assay is preferably performed on serum, both for TG and for anti-TG Ac (this is the case in our laboratory). Venous blood is collected in a dry, anticoagulant-free tube. Alternatively, the assay can be performed on plasma; in this case, the blood sample is collected in a tube with an anticoagulant: EDTA (ethylene diamine tetra-acetic acid), heparin (valid for TG and anti-TG assays) or citrate (valid for anti-TG assays).

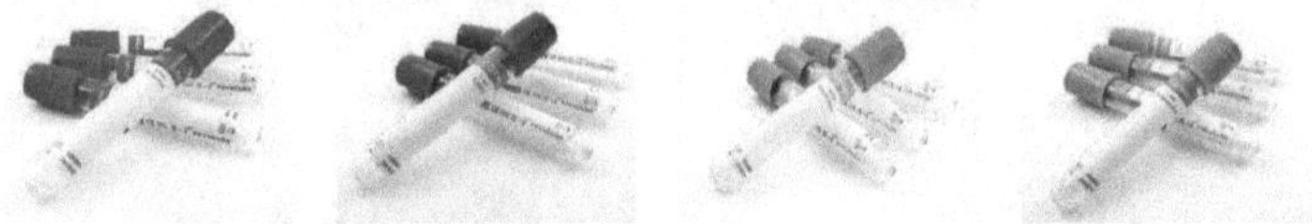

Figure 6: Sampling tubes

Red cap: tube without anticoagulant Purple cap:
tube with anticoagulant (EDTA)
Green cap: tube with anticoagulant (Heparin) Blue cap:
tube with anticoagulant (Citrate)

- Correct identification of the sample is imperative. The label to be affixed to the tube must contain at least the following information:

- Patient's full name
- Prescribing department
- Sample date

- The identified sample is attached to the prescription sheet (or laboratory analysis request). This must include the following information

- Unambiguous patient identification (first and last name, registration number or file number)
- Prescribing physician / department
- Tests requested
- Patient's clinical details, including sex and age For TG assays, it must be specified whether the sample was taken during TSH braking (patient on L-thyroxine) or after TSH stimulation (1 month after L-thyroxine discontinuation).
- Sample date

➢ **Routing :**

The preceding steps are carried out outside our laboratory: in a clinical department at the patient's bedside (for inpatients), in the hospital sampling unit (for outpatients) or elsewhere.

Generally, no special transport conditions are required; the whole blood sample is transported to the laboratory on the same day (< 6 hours) at room temperature (between 15 and 25°C).

➢ **Reception :**

On arrival at the laboratory, samples are sorted according to the analysis requested and distributed to the corresponding bench.

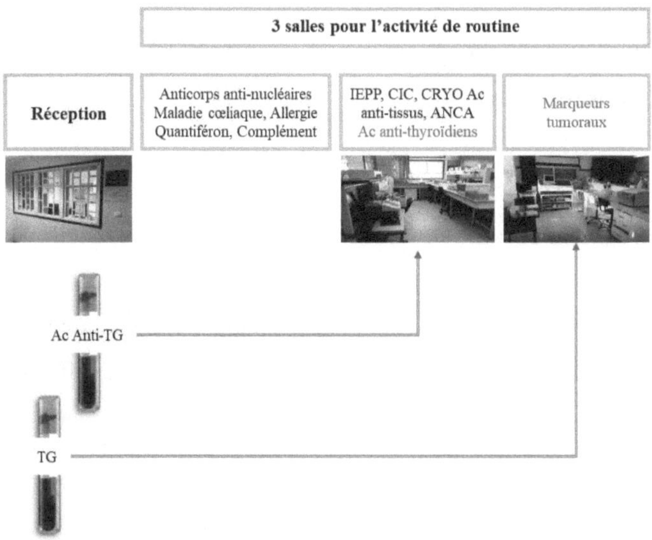

Figure 7: Sample routing within the laboratory

At the lab bench, you must :

- Check that the sample is correctly identified: "sample-request" pair.
- Assign a unique order number to each tube-request
- Detect non-conformities (e.g. identity discrepancies, missing tube identification, missing or incomplete prescription sheet, unsuitable tube, etc.).

→ Sample accepted, accepted "with reservation" or refused

- **Recording :**

Each sample received is recorded in the bench register (Excel file entry with printed version for TG, conventional register entry for anti-TG Ac). The data recorded for each sample includes

- Date received
- Order number
- Patient's first and last name
- Prescribing department
- Patient code (for bar-coded samples) and/or Personnel number and/or File number
- Clinical information
- Analysis requested

- **Pre-treatment :**

✓ **Centrifugation of samples**

Principle

Centrifugation is a process for separating compounds in a mixture according to their density difference by subjecting them to centrifugal force.

The apparatus used is a high-speed rotating machine known as a centrifuge.

Figure 8: Centrifuge

In the case of whole blood collected from a dry tube, centrifugation, after complete coagulation, produces a solid phase (compact aggregate of fibrin and blood cells) and a liquid phase, the supernatant (serum).

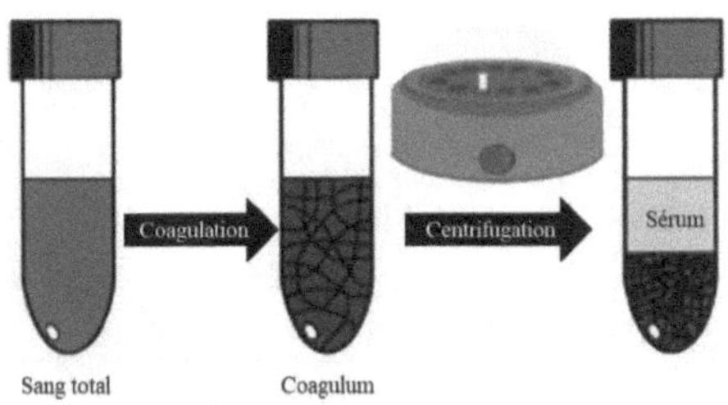

Figure 9: Serum preparation

Procedure

- Lift centrifuge hood.
- Place tubes in the appropriate buckets in a balanced manner.
- Lower and lock the lid; a safety device prevents starting if the lid is incorrectly locked.
- Select centrifugation speed in rpm (3000 rpm) and time in minutes (10 min).
- Press the start button.

- When the centrifuging time has elapsed, the unit stops automatically and emits an audible signal.

- Unlock after stop stop ; a device of safety device prevents the lid from opening during centrifugation.

- Lift lid.

- Remove tubes.

✓ **Decanting and aliquoting**

The serum obtained after centrifugation is collected with a micropipette from the primary sample tube and transferred to a 5 ml hemolysis tube (for analysis) and an Eppendorf (for reserve/serum library). These secondary samples are identified by the same serial number as the primary sample. Ideally, an additional barcode label is supplied to be affixed to the tube intended for analysis (particularly for automated assays with barcode reader).

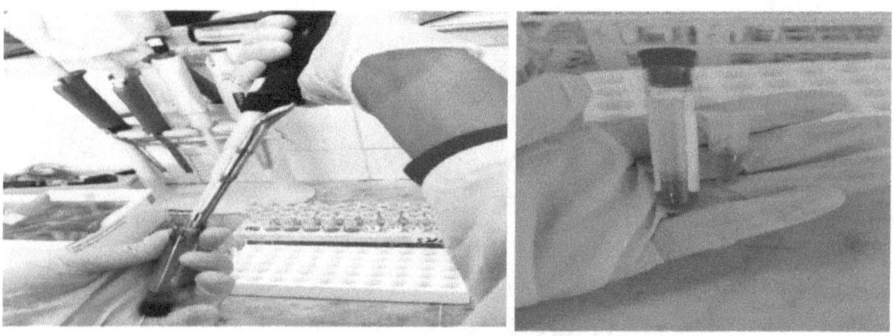

Figure 10: Decanting and aliquoting

Primary sample tubes are deposited a suitable container and join the circuit of waste from infectious risk sanitary activities.

➤ **Storage :**

For TG assay, samples are stable for :

- 48 hours at +15°C to +25°C
- 72 hours at +2°C to +8°C

- 1 month at -20°C

For anti-TG Ac assays, samples are stable for 14 days at +2°C to +8°C.

2. Analytical phase :

2.1. Determination of thyroglobulin

➢ **Assay technique** :

ECL : Electro-Chimie-Luminescence on Cobas e411 (Roche)®

➢ **Principle :**

Automated immunological technique with luminescent labeling

The principle is that of the Sandwich reaction. Total analysis time: 18 minutes.

✓ **1st incubation: Immunological reaction**

35µl sample containing TG

+ R1: TG-specific monoclonal Ac, conjugated to biotin

+ R2: TG-specific monoclonal Ac, conjugated to ruthenium

➔ Formation of a sandwich complex

✓ **2nd incubation: Binding to solid phase**

+ M: streptavidin-coated microparticles

➔ Binding of sandwich complex to solid phase (microparticles) by interaction of biotin and streptavidin.

✓ **Electrochemical-luminescent reaction**:

The reaction mixture is drawn into the measuring cell, where the microparticles are magnetically attracted to the electrode surface.

Unbound substances are then removed with ProCell (containing TPA: Tripropylamine).

Applying a voltage to the electrode initiates a chemical reaction involving TPA and ruthenium resulting in a luminescent emission which is measured by a photomultiplier.

✓ **Determination of results :**

The luminescent signals are converted into concentrations using a calibration curve (two-point calibration).

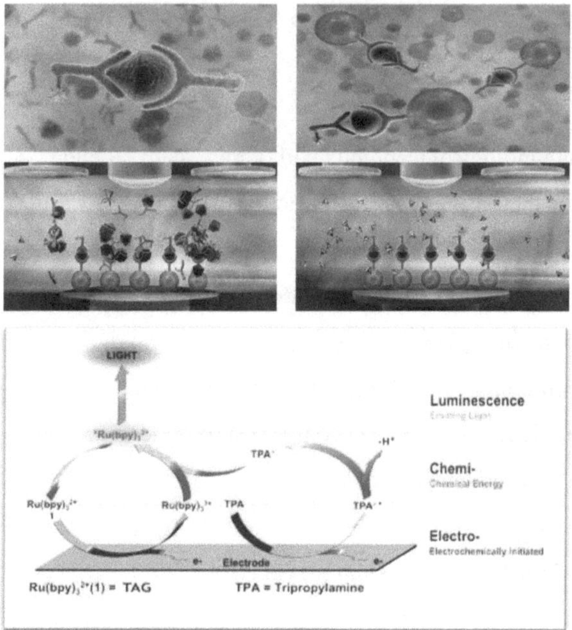

Figure 11: Principle of electro-chimiluminescence

➢ **Overview of the analyzer :**

The Cobas e411 analyzer is a fully automated instrument controlled by a control unit via user software. It comprises 3 zones:

- Sample/reagent area

- Sample disk: 30 positions→ samples, controls, calibrators
- Reagent rotor: 18 positions→ reagents (M, R1, R2), pre-treatments or diluents.

- <u>Consumables area</u>

- 3 tip racks
- 3 cuvette racks
- Pipetting station: 5 positions→ tips, cuvettes for sample dilution or pre-treatment, reaction cuvette
- System water tank (deionized or distilled water+ Elecsys Sys Wash)
- Liquid waste container
- Solid waste container

- <u>Measurement area</u>

- Incubator: 32 positions→ reaction cuvette (reaction mixture maintained at 37°C)
- Suction station
- System reagents: ProCell: washing solution, buffer containing TPA
  CleanCell: measuring cell cleaning solution
- Detector

Figure 12: External view of the Cobas e411 controller

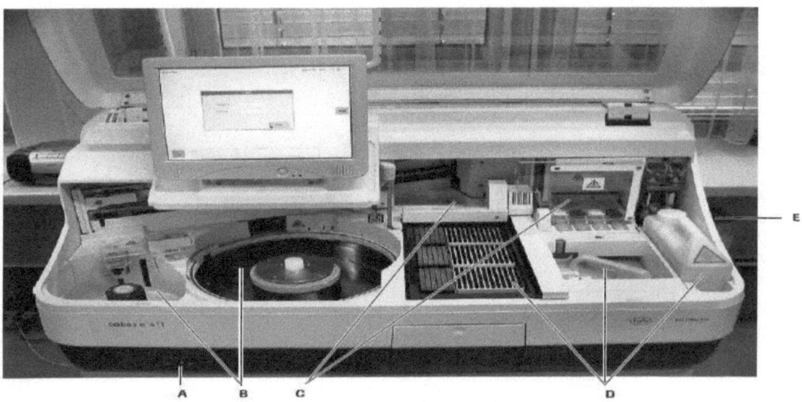

Figure 13: Internal view of Cobas e411 controller

A: Power switch
B: Sample and reagent rotors
C: Measurement area
D: Consumables and liquid waste area
E: Main circuit breaker (hidden)

- **Operating procedure :**

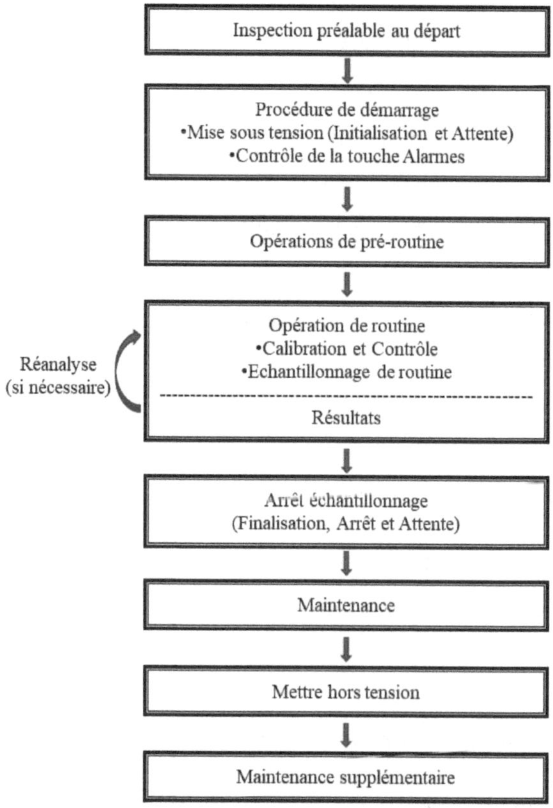

Figure 14: Flow chart of daily operations.

✓

✓ **Start-up procedures :**

- Power up
- Log on
- Check system alarms
- Troubleshoot alarms (if necessary)

✓ **Pre-routine operations:**

- Place required reagent packs
- Replace system reagents (ProCell and CleanCell) (if necessary)
- Replace cuvette and tip racks

- Fill system water tank (if necessary)
- Empty liquid and solid waste containers

✓ **Instrument calibration/quality control :**

- Check if calibration or quality control is required by the system
- Check whether calibration needs to be renewed for each reagent pack
- Prepare calibrator and control materials (if required)
- Load required calibrators and controls
- Dose calibrators and controls
- Validate calibration and control results

✓ **Sample analysis :**

- Program patient samples
- Load patient samples
- Start routine
- Monitor sample processing

2.2. Anti- thyroglobulin antibody assay

➢ **Assay technique :**

ELISA: Enzyme-Linked Immunosorbent Assay, at using a Anti-Thyroglobulin ELISA (IgG)" kit (Euroimmun)®

➢ **Principle:**

Enzyme-labelled immunoassay (Enzyme-labelled immunoassay)

In our case, this is an indirect ELISA in which the Ag (adsorbed) - Ac (to be assayed = primary Ac) reaction is revealed by a color reaction produced by the action of an enzyme, attached to a secondary Ac (recognizing the primary Ac), on its substrate.

The kit used enables quantitative determination of human IgG class TG antibodies in serum or plasma.

The kit contains 12 strips, each with 8 wells. The Ag adsorbed in each well is purified native TG isolated from human thyroid.

In the first step, **diluted samples** are incubated in the wells. In the case of positive samples, specific IgG Ac (also IgA and IgM) bind to the adsorbed Ag. In a second step, the bound IgG-isotype Ac is detected by an **enzyme-labelled anti-human IgG secondary Ac**. The enzyme (peroxidase), after addition of the corresponding **chromogen/substrate** (TMB/$H_{(2)}$ O2), catalyzes a color reaction:

H_2O_2+ [colorless TMB]-H_2→ 2 H_2O+ [oxidized TMB=blue-colored product]

*TMB: 3,3′, 5,5′-tetramethylbenzidine

The intensity of the color formed is proportional to the concentration of anti-TG IgG Ac.

The reaction is stopped by the addition of an acid buffer (sulphuric acid) inhibiting enzymatic activity, inducing a turn from blue to yellow.

The **optical density (OD)** of each sample is **measured spectrophotometrically**, then **converted to concentration** using a calibration curve (3-point calibration).

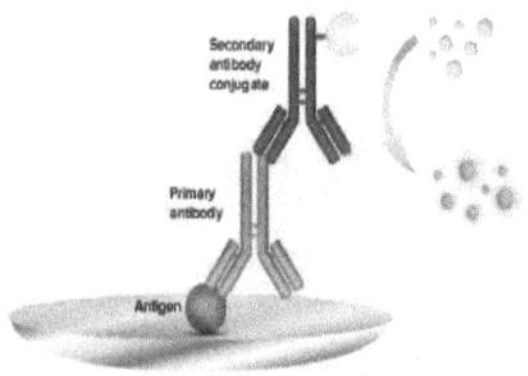

Figure 15: Indirect ELISA

➢ **Procedure :**

✓ **Preparation of pipetting protocol :**

	1	2	3	4	5	6	7	8	9	10	11	12
A	C1	S4										
B	C2	S5										
C	C3	S6										
D	T+	S7										
E	T-	S8										
F	S1	S9										
G	S2	S10										
H	S3	S11										

Figure 16: Pipetting protocol

C1: Calibrator 1 C2: Calibrator 2 C3: Calibrator 3

T+: Positive control T-: Negative control S1→ S11: Patient sera

✓ **Reagent preparation :**

- Bring kit reagents to room temperature (+18°C to +25°C)

  The calibrators, controls, buffer sample buffer, conjugate, chromogen/substrate, stop solution are ready to use.

  Wash buffer (10-fold concentrated) should be diluted i n distilled water.

✓ **Sample dilution** (1/200)**:**

- Prepare sample tubes (sera) for analysis
- Prepare empty hemolysis tubes and number them in the same way as the samples to be analyzed
- Dispense 1 ml of sample buffer into each tube using a micropipette dispenser.
- Add 5 µl of each serum to the corresponding hemolysis tube

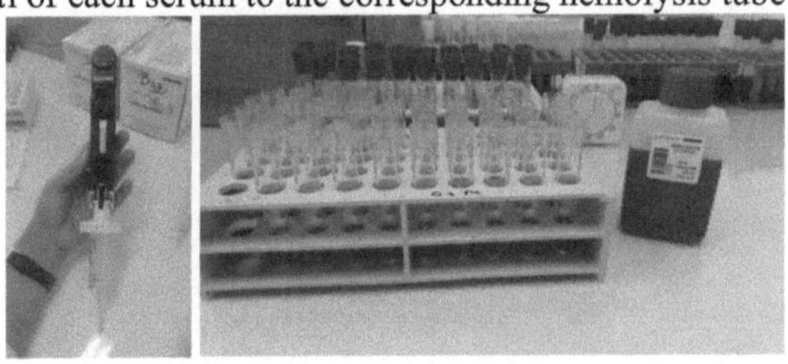

Figure 17: Serum dilution

✓ **Sample incubation :**

- Transfer 100 µl of calibrators, positive and negative controls and diluted sera to microplate wells according to pipetting protocol.
- Incubate for 30 minutes at room temperature

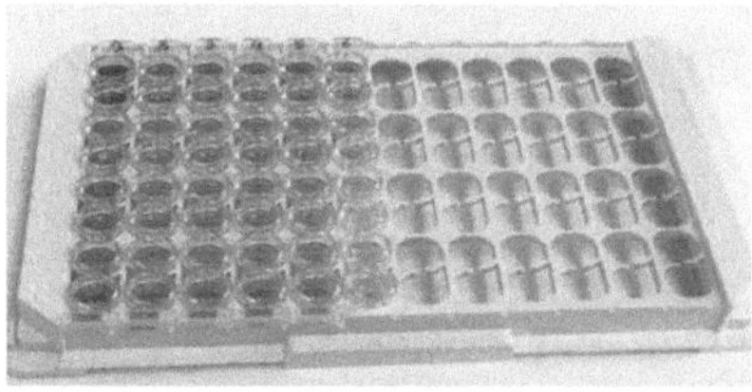

Figure 18: Microplate incubated with samples

- ✓ **Washing :**
  - o Prepare wash solution: 450 ml distilled water + 50 ml wash buffer
  - o Use the automatic washer to wash the wells 3 times with 400 µl of wash solution.

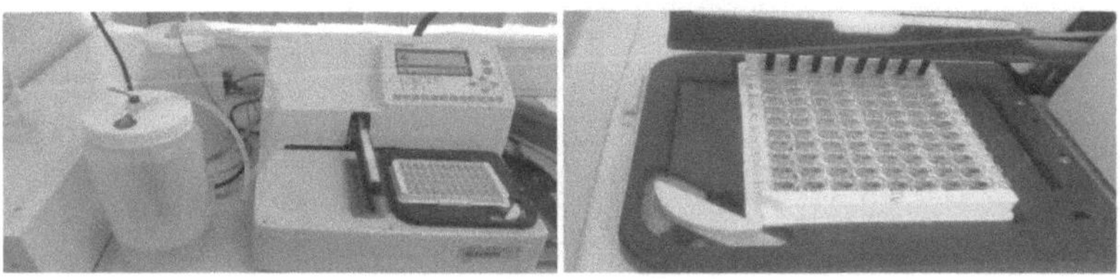

Figure 19: Automatic ELISA microplate washer

- ✓ **Incubation with the conjugate (Ac anti-IgG human labelled à conjugate):**
  - o Add 100 µl of conjugate to each microplate well using a multichannel propette.
  - o Incubate for 30 minutes at room temperature

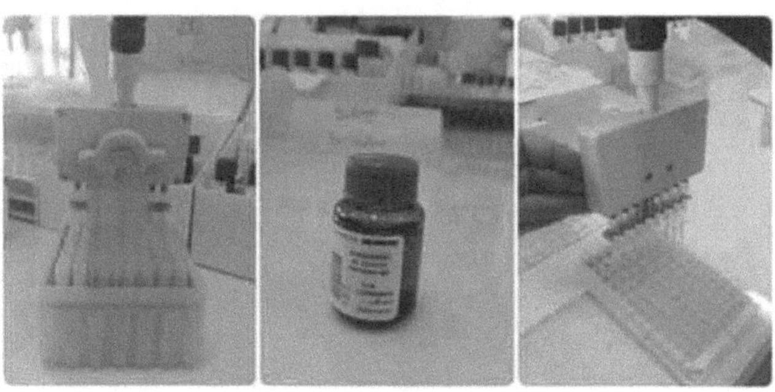

Figure 20: Addition of conjugate

✓ **Washing**: as described above

✓ **Incubation with chromogen/substrate :**

- Add 100 µl of chromogen/substrate to each well from microplate using a multichannel micropipette
- Incubate for 15 minutes at room temperature in the dark

No washing after this step.

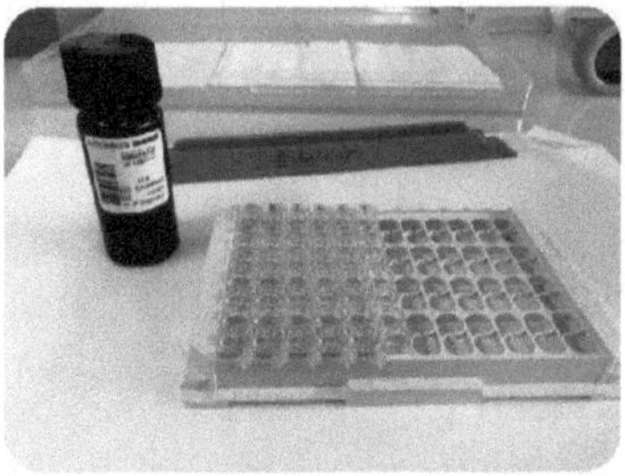

Figure 21: Adding chromogen/substrate

✓ **Stop reaction :**

- Add 100 µl of stop solution to each microplate well using a multichannel propette.

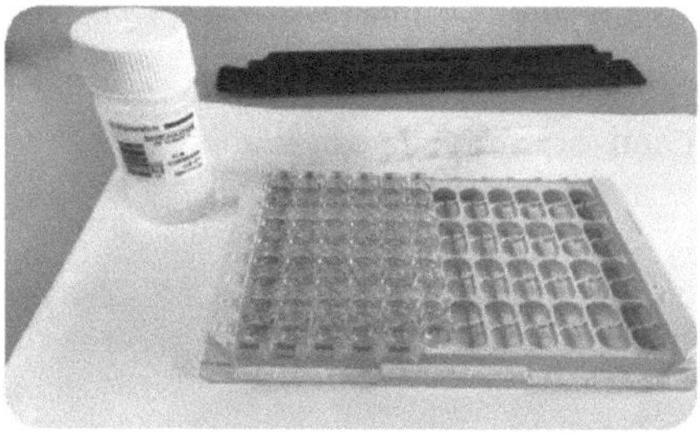

Figure 22: Adding stop solution

✓ **Spectrophotometer measurement:** within 30 minutes of adding stop solution

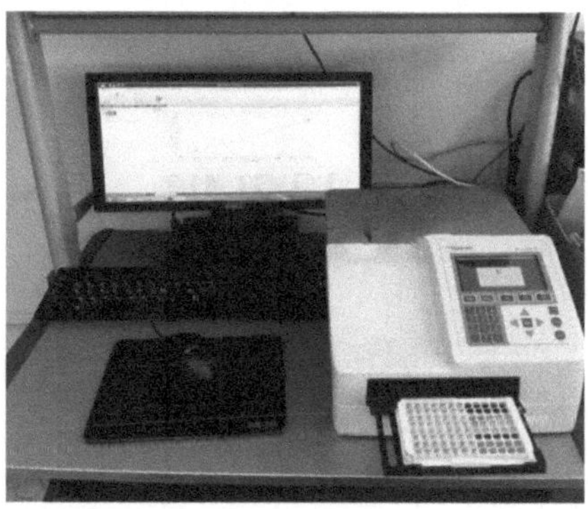

Figure 23: Spectrophotometer reading

3. Post-analytical phase :

➢ **For TG :**

Calibration and controls are carried out and validated before patient samples are processed.

TG concentration is expressed in ng/ml.

- **For anti-TG Ac :**

Calibration and controls are carried out in parallel with patient samples.

To validate the reaction, please refer to the quality control certificate supplied with the kit.

The concentration of anti-TG antibodies is expressed in IU/ml (normal values <100 IU/ml).

IV. Results and Interpretation

During the study period (2 months), 35 requests for serum TG assays were included. This assay was indicated in these patients as part of the follow-up of differentiated thyroid cancer treated by total thyroidectomy (followed by isotopic destruction of remnants with iodine 131). These requests came mainly from the Nuclear Medicine department (97.1%).

Table 1: Characteristics of cases included in the study

Patient	Department	Sex	Age (years)	Mean follow-up time after thyroidectomy (years)
1	MN	F	60	10
2	MN	F	56	5
3	MN	F	48	6
4	MN	F	47	2
5	MN	F	49	6
6	MN	F	33	9 months
7	MN	F	70	-
8	MN	F	57	9 months
9	MN	F	68	3 months
10	MN	F	64	11
11	MN	F	52	-
12	MN	F	45	13
13	MN	F	51	8 months
14	MN	F	31	9 months
15	MN	F	21	1
16	MN	F	50	9 months
17	MN	F	60	1
18	MN	H	30	2
19	MN	F	44	9 months
20	MN	F	38	1
21	MN	F	61	11 months
22	MN	F	40	11 months
23	ENT	H	63	3
24	MN	F	-	3 months
25	MN	F	62	3
26	MN	H	40	6 months
27	MN	F	31	6
28	MN	F	37	3
29	MN	F	53	2
30	MN	F	18	3 months
31	MN	F	43	17
32	MN	F	45	2
33	MN	H	46	10
34	MN	F	44	2
35	MN	F	63	2

MN: Nuclear Medicine; ENT: Ear, Nose and Throat; **W:** Female; M: Male

Table 2: Results of TG and anti-TG Ac assays

Patient	TG result (ng/ml)	Anti-TG Ac result (IU/ml)	Interpretation*
1	<0,04	<20	Result in favor of an excellent response
2	<0,04	<20	Result in favor of an excellent response
3	<0,04	<20	Result in favor of an excellent response
4	0,12	<20	Result in favor of an excellent response
5	<0,04	<20	Result in favor of an excellent response
6	<0,04	<20	Result in favor of an excellent response
7	<0,04	197	Positive anti-TG antibodies, TG result not interpretable
8	<0,04	103	Positive anti-TG, TG result not interpretable
9	0,211	<20	Result in favor of an excellent response
10	1,6	<20	Result in favor of an incomplete response
11	0,422	<20	Result in favor of an excellent response
12	0,131	<20	Result in favor of an excellent response
13	<0,04	20,6	Result in favor of an excellent response
14	19,65	<20	Result in favor of an incomplete response
15	<0,04	<20	Result in favor of an excellent response
16	<0,04	<20	Result in favor of an excellent response
17	7,78	<20	Result in favor of an intermediate response
18	<0,04	<20	Result in favor of an excellent response
19	0,975	<20	Result in favor of an excellent response
20	44,27	<20	Result in favor of an incomplete response
21	33,55	27,2	Result in favor of an incomplete response
22	<0,04	<20	Result in favor of an excellent response
23	<0,04	<20	Result in favor of an excellent response
24	<0,04	23,2	Result in favor of an excellent response
25	<0,04	<20	Result in favor of an excellent response
26	1,05	<20	Result in favor of an incomplete response
27	0,516	<20	Result in favor of an excellent response
28	<0,04	<20	Result in favor of an excellent response
29	1,53	<20	Result in favor of an intermediate response
30	51,32	<20	Result in favor of an incomplete response
31	0,373	<20	Excellent response
32	0,784	<20	Result in favor of an excellent response
33	11	<20	Result in favor of an incomplete response
34	<0,04	<20	Result in favor of an excellent response
35	<0,04	<20	Result in favor of an excellent response

TG: thyroglobulin; **Ac:** antibodies

* see appendix

TG measured during TSH braking

TG measured after TSH stimulation

1- Gender distribution :

- There was a clear predominance of women (31 women versus 4 men).

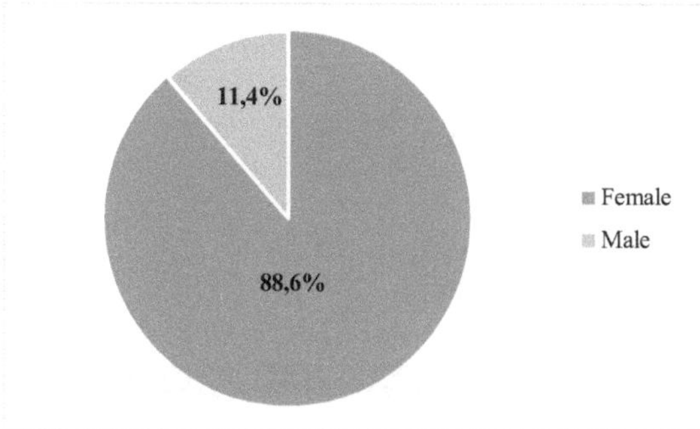

Figure 24: Distribution of patients by gender

2- Age distribution :

- The average age of patients at the time of sampling was 46.28 years (between 18 and 70 years).

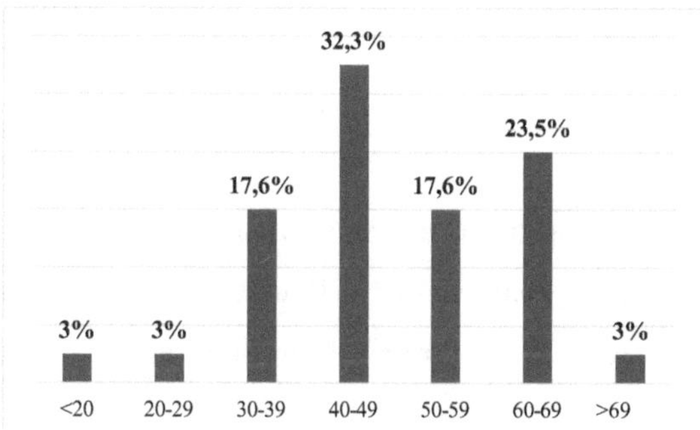

Figure 25: Age distribution of patients

→ *The demographic characteristics of the patients included in our study are consistent with data in the literature.*

3- Distribution according to length of follow-up after thyroidectomy :

- The mean duration of follow-up at the time of sampling was 3.19 years (3 months- 17 years).

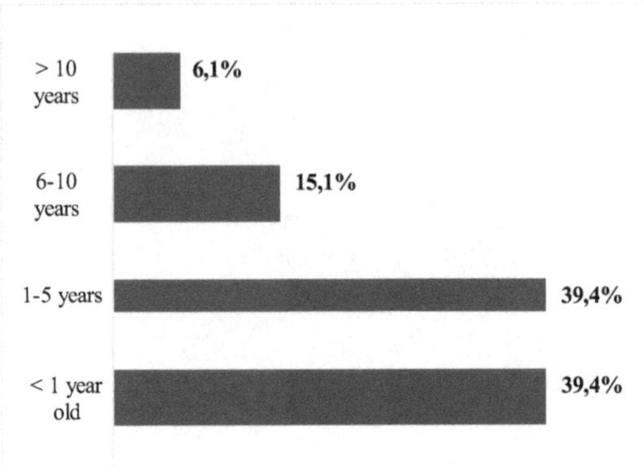

Figure 26: Distribution of patients according to length of follow-up after thyroidectomy

➔ *The overall prognosis for differentiated thyroid cancer is favorable. Only 5% of patients die of their cancer; 10-20% will relapse. Late relapse is possible, requiring lifelong surveillance.*

4- Distribution according to TG concentration :

- TG was undetectable (<0.04 ng/ml) in 18 cases. In the other cases (detectable TG), the concentration ranged from 0.12 to 51.32 ng/ml.

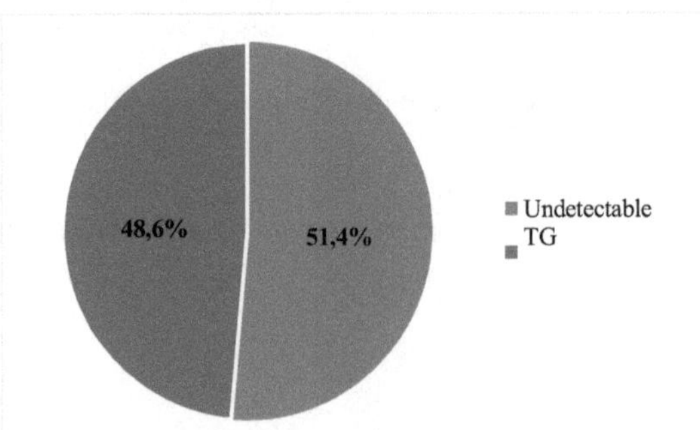

Figure 27: Distribution of patients according to TG concentration

→ *There is no "normal value" for a thyroidectomy patient! Total thyroidectomy patients should have an undetectable TG level in their serum, even if TSH is elevated.*

5- Distribution according to anti-TG Ac results :

- Anti-TG Ac was positive (>100 IU/ml) in 2 cases. In both cases, the TG concentration was <0.04 ng/ml; however, the TG result cannot be interpreted.

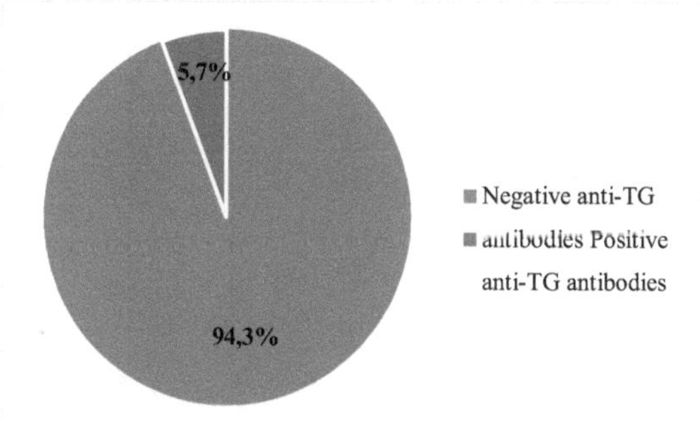

Figure 28: Distribution of patients by anti-TG antibody result

→ *The prevalence of anti-TG antibodies is higher in patients with differentiated thyroid cancer than in the general population. These Ac bind to circulating TG and can interfere with its assay (possible false negative or underestimation). In this context, an undetectable TG level does not necessarily mean remission, and a detectable TG level should be regarded as minimal. In the vast majority of cases, anti-TG antibodies disappear within a year of initial treatment. Their persistence at high titre suggests the existence of residual disease. Follow-up with anti-TG antibodies therefore represents an alternative to TG testing in these patients.*

6- Distribution of anti-TG negative patients according to T G result:

In the absence of anti-TG Ac (33 cases), the TG result was interpreted according to whether the assay was performed :

- during TSH braking (15 cases)
- or after TSH stimulation (18 cases)

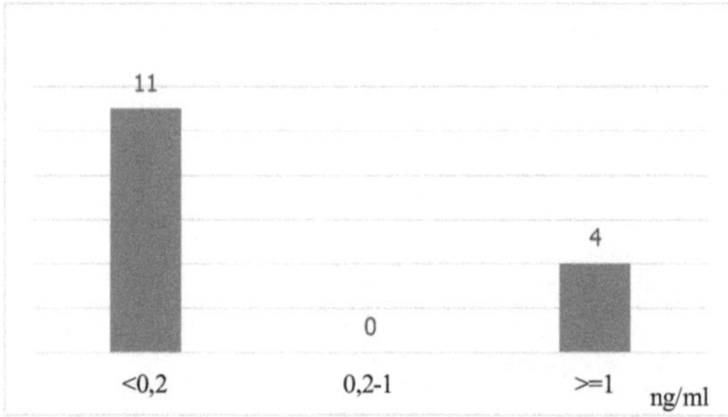

Figure 29: Distribution of patients who had a TG assay during TSH braking

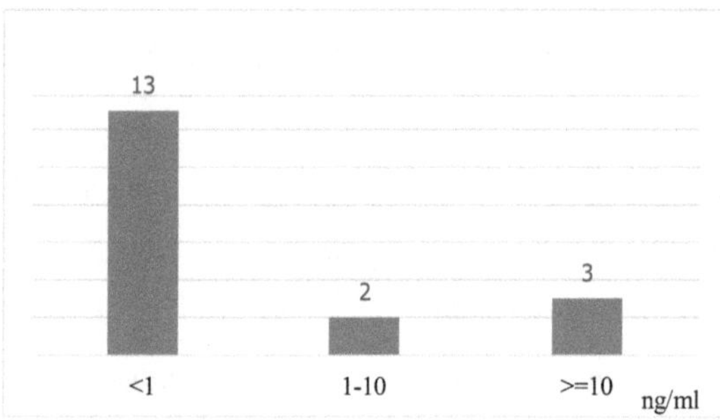

Figure 30: Distribution of patients who had a TG assay after TSH stimulation

The TG result indicated an excellent response in 24 cases, an intermediate response in 2 cases and an incomplete response (residual disease or relapse) in 7 cases.

➔ *When monitoring patients undergoing thyroidectomy for differentiated thyroid cancer, serum or plasma TG concentrations must always be interpreted in the light of the presence or absence of anti-TG Ac and TSH levels.*

TG levels can be used to assess therapeutic efficacy and detect residual disease or relapse. However, these results must always be compared with other clinical and imaging data to adapt the therapeutic and monitoring protocol.

V. Conclusion

Differentiated thyroid cancer accounts for over 85% of thyroid cancers, thyroid cancer being (although relatively rare, at around 1% of cancers) the most common endocrine organ cancer encountered clinically.

The management of patients with differentiated thyroid cancer is multidisciplinary (ENT surgery, nuclear medicine, endocrinology, etc.). After thyroidectomy, TG is a useful marker for monitoring these patients. Its measurement must be combined with anti-TG antibodies to eliminate any possible interference.

In this context, we were interested in the assay of these two parameters, namely TG and anti-TG Ac. We therefore conducted a prospective study over a 2-month period, including patients treated for differentiated thyroid cancer for whom a request for TG assay was sent to the Immunology Laboratory (CHU Habib Bourguiba, Sfax). The serum TG assay was performed by electro-chimiluminescence on the Cobas e411 (Roche®). Serum anti-TG antibodies were determined by ELISA (Euroimmun).®

During the study period, 35 patients were included. Demographic characteristics (mean age 46.28 years, predominantly female) were consistent with the literature. Mean follow-up time at the time of sampling was 3.19 years.

TG was undetectable in 18 cases and detectable in 17 cases (serum concentration between 0.12 and 51.32 ng/ml). The TG result could not be interpreted in 2 cases, due to the positivity of anti-TG Ac. In the remaining 33 cases, the TG result was interpreted according to whether the assay was

performed during TSH braking (15 cases) or after TSH stimulation (18 cases), and was in favor of an excellent response in 24 cases, an intermediate response in 2 cases, and an inadequate response in 7 cases.
However, the results of these biological analyses must be integrated with the rest of the clinical and paraclinical data to guide patient management.

Bibliography

1. Schlumberger M, Berdelou A. Thyroid cancer. La ligue contre le cancer. 2017;

2. Schlumberger M. Papillary and follicular thyroid cancer. Ann Endocrinol. 2007;68:120-8.

3. Krahenbuhl T, Portmann L, Anchisi S. Differentiated thyroid cancers: management and place of new targeted therapies. Rev Med Suisse. 2012;8:1112-7.

4. Herbomez M. Exploration biologique de la thyroïde. Rev Franc des Lab. 2009;411:39-44.

5. Taïeb D, Lussato D, Zotian E, Mundler O. Thyroglobulin and differentiated thyroid cancers: clinical applications. Médecine Nucléaire. 2006;30(1):37-40.

6. D'Herbomez M, Lion G, Béron A, Wémeau J, DoCao C. Advances in thyroglobulin assays and their impact in the management of differentiated thyroid cancers. Ann Biol Clin. 2016;74(1):21-7.

7. Doggui R. Immunoanalytical profile of anti-thyroglobulin antibodies. Ann Biol Clin. 2018;76(6):695-704.

8. Haugen BR, Alexander EK, Bible KC, Doherty GM, Mandel SJ, Nikiforov YE, et al. 2015 American Thyroid Association Management Guidelines for Adult Patients with Thyroid Nodules and Differentiated Thyroid Cancer. Thyroid. 2016;26(1):1-133.

Appendix

Category	Definitions
Excellent response	Negative imaging **AND** - TG during TSH braking <0.2 ng/ml - or TG after TSH stimulation <1 ng/ml
Incomplete biological response	Negative imaging **AND** - TG during TSH braking ≥1 ng/ml - or TG after TSH stimulation ≥10 ng/ml - or elevation of anti-TG antibodies
Morphological response incomplete	Obvious imaging abnormalities regardless of TG level with or without anti-TG Ac
Intermediate response	Non-specific imaging abnormalities or negative imaging AND - TG during TSH braking ≥0.2 ng/ml but <1 ng/ml - or TG after TSH stimulation ≥1 ng/ml but <10 ng/ml - or stable or decreasing anti-TG Ac levels

**Adapted from the American Thyroid Association recommendations for the management of patients with thyroid nodules and differentiated thyroid cancer (2015).*

Summary

Monitoring of patients treated for differentiated thyroid cancer (DTC) includes serum thyroglobulin (TG) and anti-TG antibodies (Ac). The aim of our study was to describe the methodology and interpret the results of these two parameters.

TG and anti-TG Ac assays were performed by electrochemiluminescence and ELISA, respectively.

Thirty-five patients were included. TG was undetectable in 18 cases and detectable in 17 (0.12 to 51.32 ng/ml). The TG result was not interpretable in 2 cases due to the positivity of anti-TG Ac (>100 IU/ml). In the absence of anti-TG Ac (33 cases), the TG result was interpreted according to whether the assay was performed during TSH braking (15 cases) or after TSH stimulation (18 cases), and was in favor of an excellent response in 24 cases, an intermediate response in 2 cases, and an inadequate response in 7 cases.

TG assay is useful for monitoring CDT. Anti-TG autoantibodies are recommended, as they interfere with the TG assay. The results of these analyses must be compared with clinical and imaging data to adapt management.

FSC
www.fsc.org
MIX
Papier | Fördert gute Waldnutzung
FSC® C083411

Kolibri 360 GmbH
Ferdinand-Jühlke-Straße 7
99095 Erfurt, Deutschland
produktsicherheit@kolibri360.de

Druck:
CPI Druckdienstleistungen GmbH
im Auftrag der
Kolibri 360 GmbH
Ferdinand-Jühlke-Str. 7
99095 Erfurt